I0494057

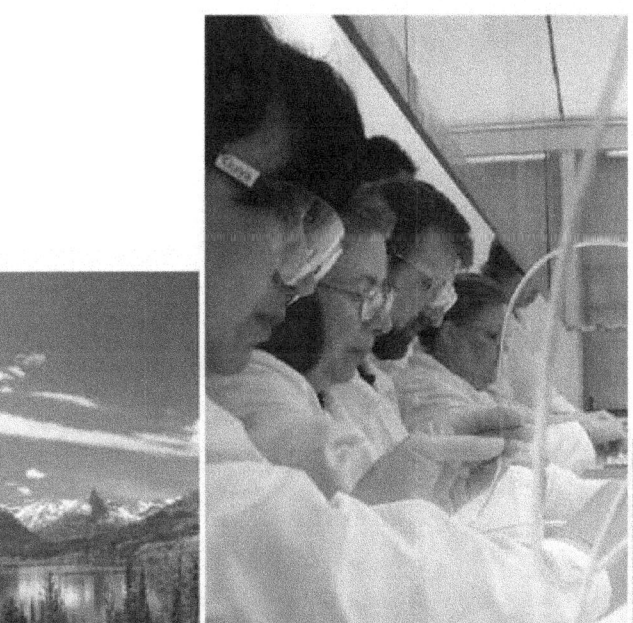

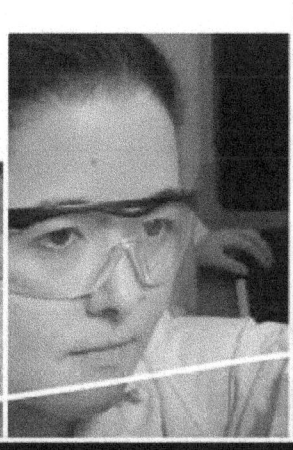
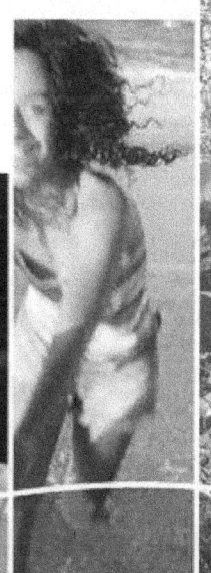

Division of
Laboratory Sciences

U.S. Department of Health and Human Services
Centers for Disease Control and Prevention
National Center for Environmental Health

CDC
SAFER · HEALTHIER · PEOPLE™

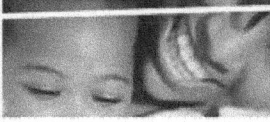
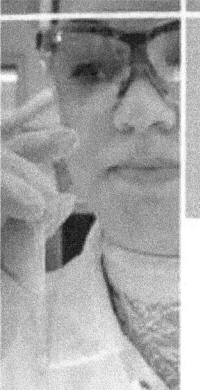

Division of Laboratory Sciences

U.S. Department of Health and Human Services
Centers for Disease Control and Prevention
National Center for Environmental Health

Centers for Disease Control and Prevention
National Center for Environmental Health
Division of Laboratory Sciences
Atlanta, Georgia 30341-3724

October 2008
NCEH Pub. No. 4-1-08-57f94

Table of Contents

Introduction

In the first decade of this new century, scientists at the Centers for Disease Control and Prevention's (CDC's) Division of Laboratory Sciences have lots to be excited about. The results of our laboratory achievements over the last 30 years have led to significant improvements in public health. Sophisticated equipment and state-of-the-art facilities are critical to that effort, but it's the ingenuity of the Division's staff that makes the difference. The CDC and the public health community salute the quality of our work as we head into the future. This brochure offers a brief glimpse of our many and diverse programs. It explains how our laboratory is organized and summarizes how each specialized laboratory in our organization uses its expertise.

We are known for our responsiveness in public health emergencies and for providing key information to guide epidemiologists in the field. We're also working in concert with state public health laboratories, providing training, proficiency testing, and advanced technology so that the nation is prepared to respond to terrorist acts involving chemicals. We are also hard at work on plans for a similar effort in response to radiologic terrorism. Still other scientists are deeply involved in exploring a new frontier—proteomics (i.e., the genomic study of proteins)—and are using their knowledge to conduct toxin research that will ultimately result in faster diagnosis of many diseases.

As you will see, one large area of concentration involves biomonitoring, which is the direct measurement of chemicals, including nutritional and dietary indicators, in people's blood or urine. At this writing, our scientists have developed methods for measuring more than 450 environmental chemicals and nutritional indicators in human biologic samples. These biomonitoring measurements also provide public policy makers and regulators with the information they need to base their policy decisions on accurate exposure information rather than on estimates of exposure.

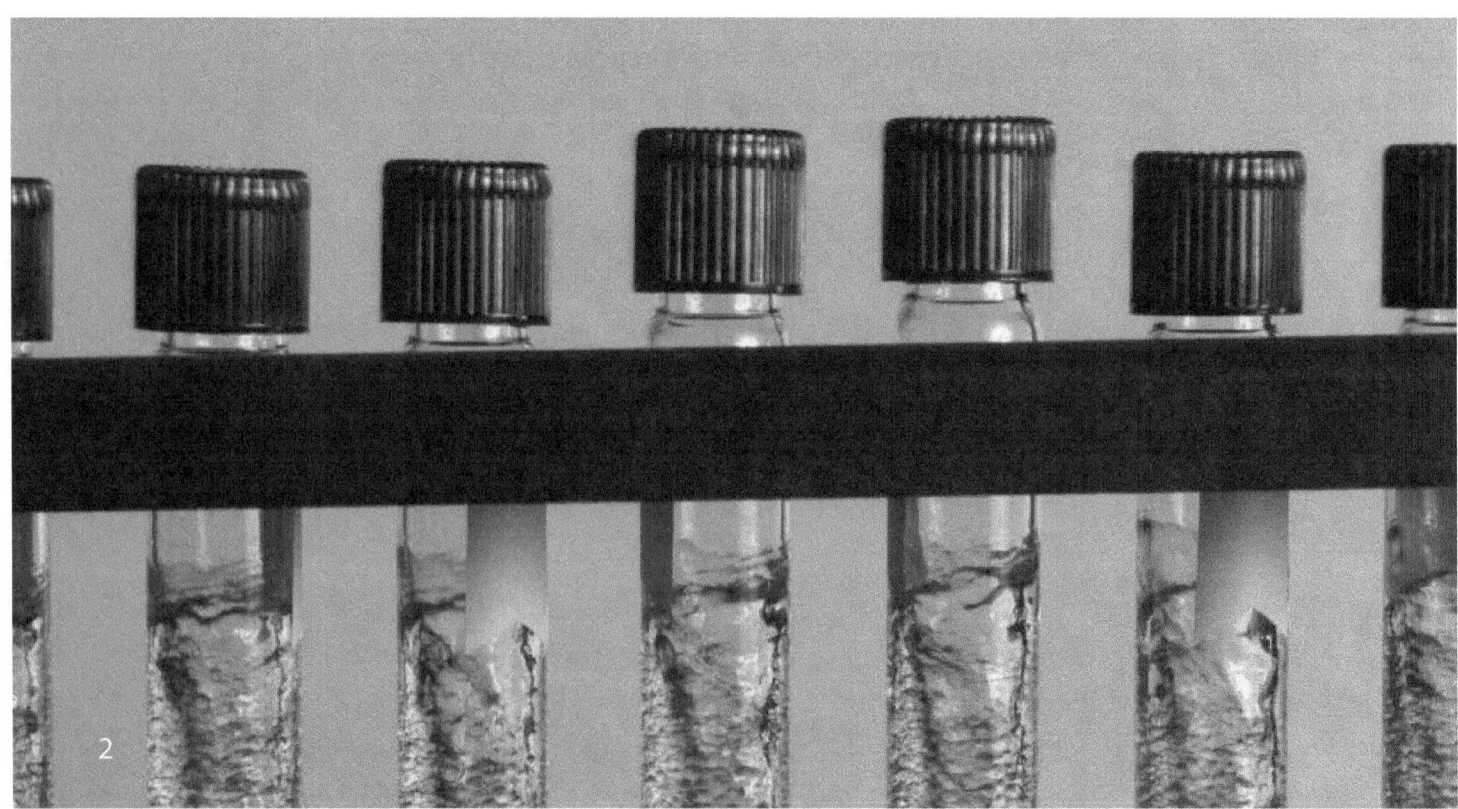

Our laboratory is also engaged in other important work. One of our long-standing programs helps assure the quality of newborn blood spot screening tests that are conducted by state public health laboratories. This program, now in its 30th year, is one of the most successful public health programs ever developed, saving several thousand children per year from dying prematurely or developing mental retardation. Other scientists are working to standardize measurements related to clinically important compounds for diagnosing, treating, and preventing chronic disease. We're also examining the relation between genetics and environmental exposures as factors in the cause of disease, and we're conducting research related to the toxic and addictive properties of tobacco products. Our laboratory is recognized as the nation's leader in tobacco research. We can measure additives and toxic substances, not only in tobacco products, but also in smoke and in people who use tobacco products or are exposed to secondhand smoke. No other laboratory in the federal government has these capabilities.

We hope this brochure gives you a glimpse of the rich diversity of our programs that allow us to approach complex public health issues from a resource and talent base that no other laboratory in the world can match. Our laboratory scientists collaborate with many partners: U.S. government agencies, state and local health departments, academic institutions, community organizations, philanthropic foundations, and international agencies and organizations. We invite inquiries about our programs as well as suggestions for scientific collaboration that would benefit public health.

Eric J. Sampson, Ph.D.
Director
Division of Laboratory Science

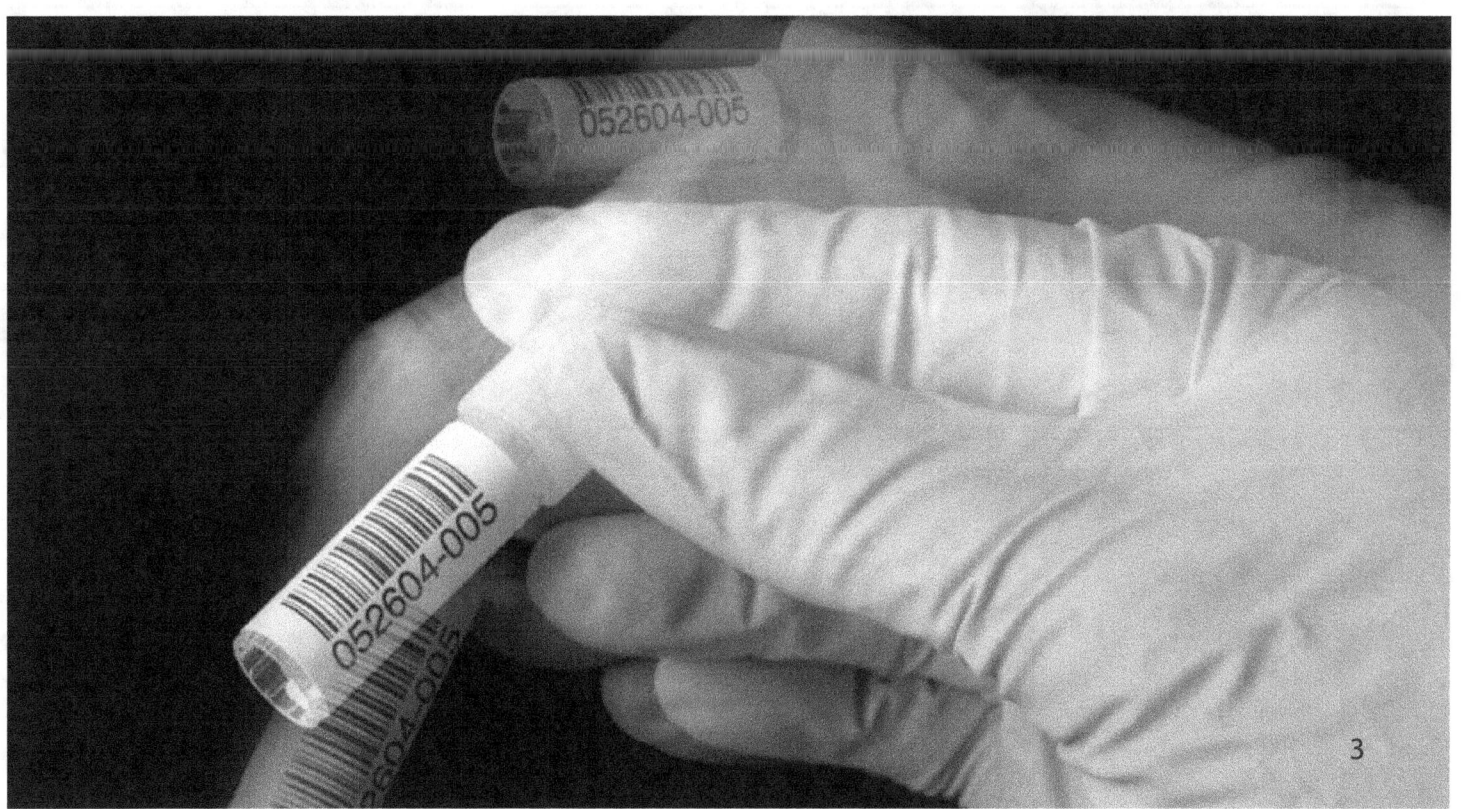

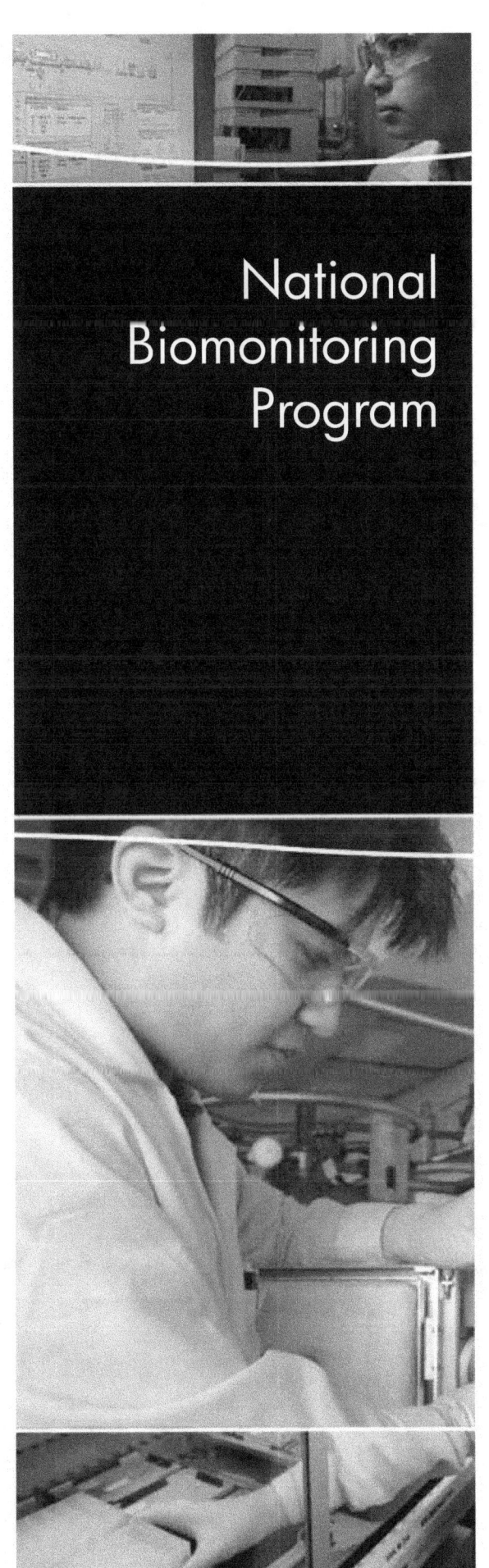

National Biomonitoring Program

For at least three decades, our scientists have been determining which environmental chemicals people have been exposed to and how much of those chemicals actually gets into their bodies. This technique is known as biomonitoring. Biomonitoring measurements are the most health-relevant assessments of exposure because they measure the amount of the chemical that actually gets into people, not the amount that *may* get into people.

Our laboratory operates CDC's National Biomonitoring Program (NBP). Throughout the world, biomonitoring has become the standard for assessing people's exposure to toxic substances as well as for responding to serious environmental public health problems. Rather than estimating how much of a substance gets into people from measured environmental concentrations, our scientists have taken out the guesswork by measuring levels of chemicals that actually *are* in people's bodies. And they do so with precision, speed, and pinpoint accuracy, measuring many chemicals in a very small amount—often a teaspoon or less—of blood or urine.

NBP currently measures more than 450 environmental chemicals and nutritional indicators in people. All of the methods used have been published in peer-reviewed journals so that other laboratories can use them. NBP also shares its methods with many state public health laboratories, and the program trains these laboratories in the use of these methods.

Each year, NBP works with many different groups, including state health departments, to provide exposure information for public health investigations or emergencies as well as for 60–70 exposure studies. Health officials need biomonitoring information to help them make the best decisions that will benefit the health of the American public. Our scientists also collaborate with U.S. government agencies, state and local health departments, universities, community organizations, and international organizations on national studies of general population exposure and studies of specific exposed populations, such as children.

National Report on Human Exposure to Environmental Chemicals

NBP publishes the *National Report on Human Exposure to Environmental Chemicals*, which is an ongoing assessment of the U.S. population's exposure to environmental chemicals. Scientists measure chemicals or their metabolites in blood and urine samples obtained from a random sample of participants in CDC's National Health and Nutrition Examination Survey (NHANES). NBP measures environmental chemicals for the population as a whole and for subgroups characterized by age, sex, and race or ethnicity. These data provide unique exposure information to scientists, physicians, and health officials. Public health officials can use the data to—

- Determine which chemicals get into people and at what concentrations.

- Determine, for chemicals with a known toxicity level, the proportion of the population with toxicity levels associated with adverse health effects.

- Establish reference ranges that can be used by physicians and scientists to determine whether a person or group has an unusually high exposure.

- Assess the effectiveness of public health efforts to reduce exposure of people to specific chemicals.

- Determine whether exposure levels are higher among minorities, children, women of childbearing age, or other vulnerable groups.

- Track, over time, trends in levels of exposure of the population.

- Set priorities for research on human health effects of exposure.

Emergency Response and Air Toxicants Branch

The Emergency Response and Air Toxicants (ERAT) Branch has three main focus areas: emergency preparedness and response, tobacco research, and exposure research related to environmental air toxicants.

Emergency Preparedness and Response

ERAT develops analytical methods to measure, in people, substances such as chemicals and toxins, which could be used in acts of terrorism. ERAT developed and performs the Rapid Toxic Screen, which is a series of tests that analyzes people's blood and urine and determines the levels of 150 chemicals likely to be used by terrorists. Chemicals that can be measured include biomarkers of nerve agents, blistering agents, cyanide-based compounds, pesticides, metals, incapacitating agents, and other chemicals that can cause significant disease or death. Results of the Rapid Toxic Screen identify which chemicals were used, who was exposed to the chemicals, and how much of a particular chemical their bodies absorbed. This information is critical to medical and public health personnel managing the care of people exposed during a terrorist event as well as managing the care of people who believe that they were exposed to a deadly chemical agent.

The Branch also works with public health laboratories in states, territories, cities, and counties (through the Laboratory Response Network) to assist in expanding their chemical laboratory capacity to prepare for and respond to chemical-terrorism incidents.

ERAT oversees a select Laboratory Response Team that is available 24 hours a day, 7 days a week to respond to a chemical-terrorism event anywhere in the country. This team supports efforts to transport clinical samples from hospitals and trauma centers to laboratories, where proper testing can be done to assess people's exposure to chemical agents.

In the event of an attack involving priority toxins (e.g., botulinum, anthrax, or ricin), early detection and accurate identification and measurement are critical to enable more effective treatment and to prevent additional exposure. ERAT has identified new mass spectrometry-based methods to provide sensitive and accurate measurements of toxins in clinical samples (e.g., blood, urine, feces) and environmental samples (e.g., milk, food, water).

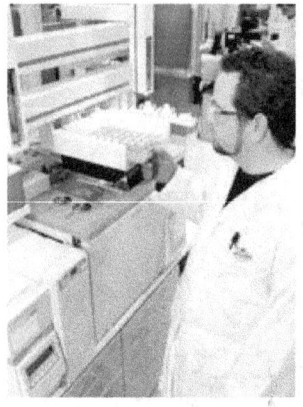

ERAT has 89 staff members, including 3 physicians, 40 people with Ph.D.s, 7 people with M.S. degrees, and 37 people with B.S. or other degrees.

Tobacco Research

The Branch conducts research, develops methods, and analyzes chemicals related to tobacco products, including tobacco itself, and tobacco smoke and its constituents. The Branch also measures, in people, levels of biomarkers related to tobacco use. ERAT has developed a method to measure very low levels of cotinine, a metabolite of nicotine and a marker of environmental tobacco smoke. Using this method, ERAT produces population-based levels for cotinine segmented by age, sex, and race or ethnicity and publishes this information in CDC's *National Report on Human Exposure to Environmental Chemicals*. Cotinine levels decreased dramatically for all segments of the population during the 1990s. The Surgeon General used ERAT's data in the 2006 Surgeon General's Report (*The Health Consequences of Involuntary Exposure to Tobacco Smoke*) to highlight the public health success in reducing nonsmokers' exposure to environmental tobacco smoke in public places. The Surgeon General's Report noted that although progress has been made to reduce environmental tobacco smoke exposure in public places, data show that 126 million U.S. nonsmokers are still exposed, and more work needs to be done to reduce this exposure.

ERAT also analyzes tobacco products and cigarette smoke for the presence of harmful substances and other substances that may influence the delivery of harmful substances to the body. For instance, the laboratory not

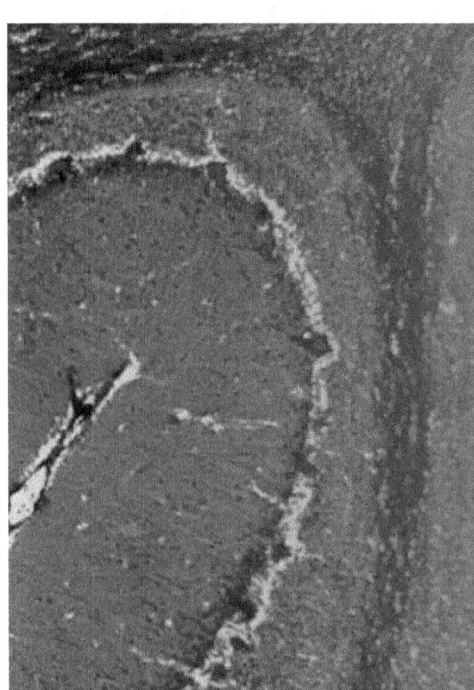

Faster Method for Measuring Botulinum Toxin in People

Botulinum toxin is a neurotoxin and is one of the most poisonous substances known. This toxic agent is near the top of the terrorism-threat list. The current test for measuring botulinum toxins requires injecting samples into mice and waiting 2 to 3 days to see if the mice die. Mice must be carefully selected and monitored continuously. Using mass spectrometry technology, CDC scientists developed a much faster method for detecting botulinum toxin in people. This method reduces testing time to about 3–4 hours, and it can identify all seven types of botulinum toxins. This method also can be used to measure botulinum toxin in environmental samples such as milk, food, and water. By providing more rapid identification of botulinum toxin, this new CDC method enables faster treatment and more rapid public health action to prevent additional exposures.

only investigates the concentration of nicotine in cigarettes and smoke but also the factors or chemicals that influence the delivery of nicotine to people. In addition to nicotine, these harmful substances include nitrosamines; carbon monoxide; volatile organic compounds (VOCs) such as benzene; metals such as cadmium and lead; carbonyl compounds; and polycyclic aromatic hydrocarbons (PAHs).

Air Toxicants Research

ERAT scientists conduct research, develop methods, and analyze samples for such environmental air toxicants as VOCs. Some of the chemicals measured are benzene, tetrachloroethene, trihalomethanes, and methyl tertiary butyl ether. Scientists in this group have also developed a method for measuring perchlorate, which is a chemical used in solid rocket propellant, explosives, pyrotechnics, flares, and other products.

Because of the Branch's expertise in measuring chemicals or toxins in people, ERAT scientists are often contacted by federal, state, local, and international governments as well as by other health organizations about laboratory issues, such as analytical methods, instrumentation, specimen preparation, quality control procedures, laboratory safety, and medical interpretation of laboratory results. In addition, Branch scientists work with investigators worldwide on studies designed to determine whether a link exists between exposure to selected chemicals and health effects.

DID YOU KNOW?

In the early 1990s, the tobacco laboratory in the ERAT Branch produced data showing that 88 percent of the nonsmoking population was exposed to tobacco smoke. This finding was used as a justification for restricting smoking in public buildings.

Measuring People's Exposure to Perchlorate

In 2006, scientists from the ERAT Branch measured the U.S. population's exposure to perchlorate, a chemical compound used in solid rocket propellant, explosives, pyrotechnics, flares, and other products. Using a new laboratory method developed by ERAT, scientists analyzed urine samples from participants in CDC's NHANES for the years 2001–2002. Researchers found measurable levels of perchlorate in the urine of all 2,820 survey participants, indicating widespread exposure to this chemical in the U.S. population. In addition, researchers examined the relation between low urinary levels of perchlorate in the U.S. population and thyroid hormone levels. Researchers found an association between levels of perchlorate in urine and decreased thyroid function in women aged 12 years old and older. The relation was strongest among women with lower iodine intake, indicated by women who had levels of iodine in their urine of less than 100 micrograms per liter. About 36 percent of women in the United States have these lower iodine levels.

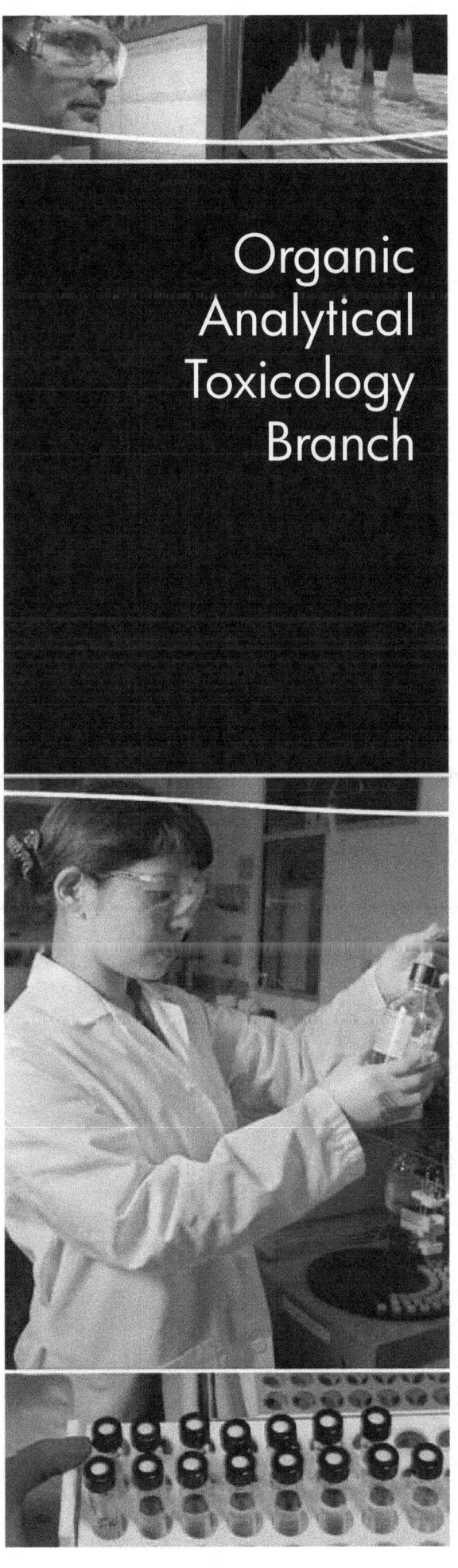

Organic Analytical Toxicology Branch

The Organic Analytical Toxicology (OAT) Branch develops analytical methods to measure synthetic and naturally occurring organic environmental chemicals in people. Among the many chemicals measured are dioxins; polybrominated biphenyl ethers; polychlorinated biphenyls; polyfluorinated alkyls; environmental phenols; PAHs; phthalates; and pesticides, including herbicides, fungicides, and insecticides. Branch scientists measure these chemicals or their metabolites in human samples (e.g., urine, blood, serum, breast milk, and meconium). For many of these chemicals, OAT produces data on population-based exposure levels segmented by age, sex, and race or ethnicity and publishes this information in CDC's *National Report on Human Exposure to Environmental Chemicals*.

OAT scientists also work with investigators worldwide on studies designed to determine whether a link exists between exposure to selected chemicals and health effects. For example, work by OAT scientists in assessing people's exposure to phthalates alerted public health officials to possible links between exposure to these chemicals and adverse health effects. In August 2008, three phthalates used in children's toys and child-care products were banned permanently, and another three were banned temporarily (pending additional extensive study) when President Bush signed the Consumer Product Safety Commission Reform Act into law.

Because Branch scientists are experts in measuring environmental chemicals in people, they are often contacted by federal, state, local, and international governments as well as health organizations about laboratory issues, such as analytical methods, instruments, specimen preparation, quality-control procedures, laboratory safety, and medical interpretation of laboratory results.

Measuring Exposure to a Chemical Found in Plastic Medical Devices

Phthalates are a group of chemicals used in making plastics. They often are called plasticizers. Phthalates can prolong the lifespan or durability of plastics and increase the flexibility of some plastics. The phthalate di-2-ethylhexyl phthalate, or DEHP, is a major component of polyvinyl chloride plastics. It often is used in medical tubing and blood-storage bags. Results of animal studies have shown that DEHP may cause reproductive health effects, specifically decreased levels of testosterone. Since 2002, OAT scientists have collaborated on several studies measuring DEHP levels in the urine of infants treated in neonatal units. The chemical can leach out of the vinyl plastic devices into patients' bodies. OAT scientists found that levels of DEHP in newborns were significantly higher than the levels measured in the overall U.S. population. As a result of these studies, several hospitals and health systems are now trying

Identifying a "Mystery Illness" in Panama

In 2006, the OAT Branch helped to figure out the cause of death of at least 50 people in the Republic of Panama. Using advanced laboratory science and innovative techniques, CDC scientists identified diethylene glycol (DEG), a chemical that is used in antifreeze, in cough syrup taken by victims of the poisoning. As a result of the Branch's work, Panamanian health authorities quickly recalled 60,000 bottles of the contaminated medications, an action that ultimately saved many lives. Following the Panama investigation, the laboratory validated the method for measuring DEG in human urine. This method was used to confirm exposure to DEG in a subsequent case-control study. In the future, the method will be a powerful tool to identify other people who have been poisoned with DEG.

The ability to measure chemicals in people is an important part of CDC's chemical-terrorism preparedness and response activities. OAT scientists work with the ERAT Branch to measure some of the chemicals in the Rapid Toxic Screen, a series of tests that analyze people's blood or urine for 150 chemicals likely to be used by terrorists. Results of the Rapid Toxic Screen will tell us which chemicals were used, who was exposed to the chemicals, and how much of a particular chemical the bodies of affected people absorbed. This information is critical to the medical and public health personnel managing the care of people in the terrorists' path, as well as to the care of people who may believe that they were exposed to a deadly chemical agent.

OAT has 69 staff members, including 18 people with Ph.D.s, 15 people with M.S. degrees, and 36 people with B.S. or other degrees.

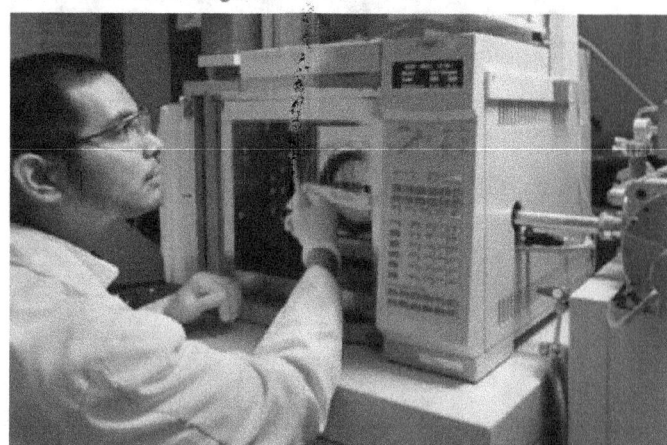

DID YOU KNOW?

The OAT Branch has developed methods to analyze at least 230 chemicals in human samples (e.g., urine, blood, serum, breast milk, and meconium).

Because of the Branch's highly advanced mass spectrometry instrumentation, many of these compounds can be measured in the parts-per-trillion range. To put this figure in perspective, one part per trillion is considered equivalent to one second in 32,000 years.

The Branch is responsible for providing analyses for approximately 90 percent of the chemicals reported in CDC's National Report on Human Exposure to Environmental Chemicals.

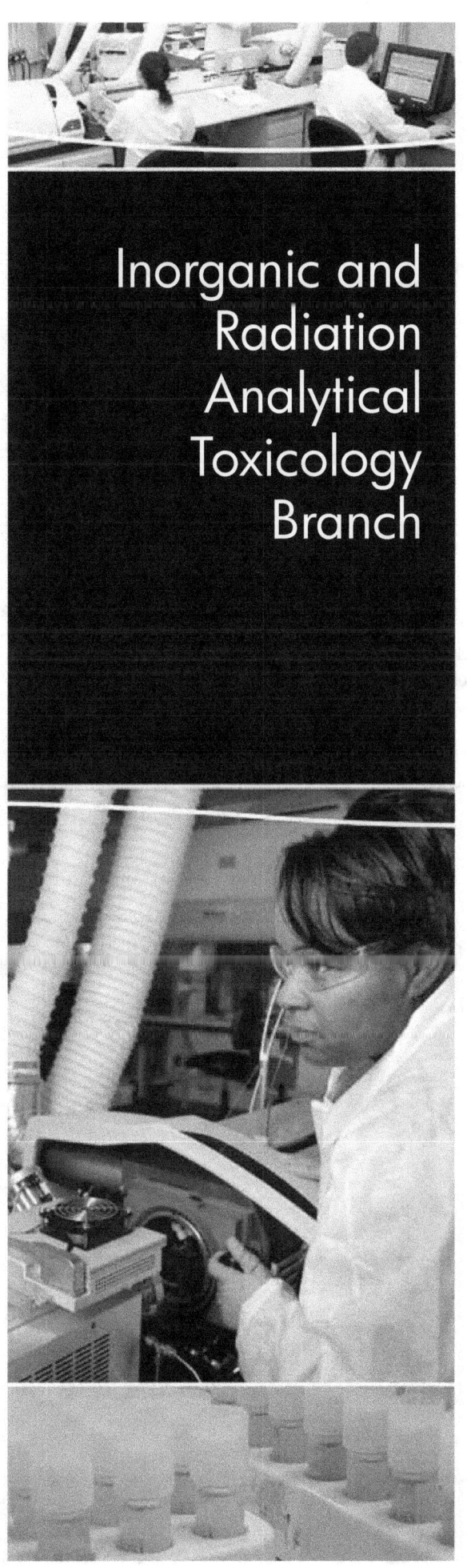

Inorganic and Radiation Analytical Toxicology Branch

The Inorganic and Radiation Analytical Toxicology (IRAT) Branch conducts research, develops methods, and analyzes elemental metals such as mercury (total and organic), arsenic (total and speciated), cadmium, lead, cobalt, tungsten, uranium, molybdenum, antimony, and other trace, toxic, and essential metals. For each of these metals, the Branch produces population-based exposure levels segmented by age, sex, and race or ethnicity. The Branch publishes this information in CDC's *National Report on Human Exposure to Environmental Chemicals*. The Branch also collaborates with academic institutions and other partners, including state public health departments, on exposure studies and studies that examine the connection between exposure levels and health effects.

Ensuring the Quality of Urinary Iodine Procedures

IRAT operates the Ensuring the Quality of Urinary Iodine Procedures (EQUIP) program, which addresses laboratory quality-assurance issues related to iodine deficiency. Iodine deficiency disorders are thought to affect more than one billion people worldwide. Accurate laboratory tests can detect iodine deficiency and, throughout the world, measuring iodine levels in people's urine is the most common way to assess the iodine status of a population. The EQUIP program currently assists more than 50 iodine laboratories in more than 30 countries, providing each laboratory with well-characterized quality-control materials, analytical guidelines, and technical training and consultation so that these laboratories can accurately measure iodine levels in their populations for national surveys. EQUIP samples are sent to participating laboratories three times each year.

Lead and Multi-element Proficiency

In 2006, IRAT launched the Lead and Multi-element Proficiency (LAMP) program, a voluntary quality-assurance program for laboratories measuring multiple analytes in whole blood. At least 100 laboratories, including 30 international labs, participate in LAMP. On a quarterly basis, these laboratories are required to analyze a set of blood samples and return test results to IRAT. IRAT provides detailed reports to the labs about their analyses. LAMP results are not used for accreditation or certification; however, the program does improve the precision and accuracy of blood lead, cadmium, and mercury measurements. In the future, IRAT will add arsenic, selenium, and uranium measurements to the program.

Laboratory Preparedness and Response to Radiologic Terrorism

IRAT also conducts research, develops methods, and performs analyses related to internal exposure (i.e., exposure in the body) to radionuclides. In the wake of a radiation event, such as the detonation of a so-called "dirty bomb," public health officials will need to 1) identify the radionuclides to which people have been exposed, 2) determine who was exposed, and 3) determine how much exposure each person received. This critical information will help guide medical treatment, prevention efforts, and other response actions. On the basis of likely terrorism scenarios, IRAT is developing methods to assess internal exposure rapidly and accurately by developing a Urine Radionuclide Screen to analyze samples for more than 20 high-priority radionuclides.

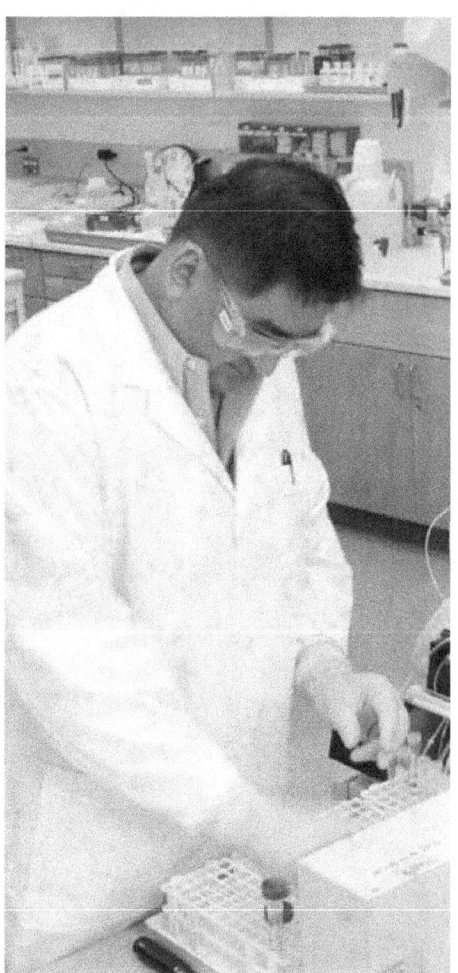

IRAT is home to a total of 34 staff members, including 9 people with Ph.D.s, 5 people with M.S. degrees, and 20 people with B.S. or other degrees.

Measuring Blood Lead Levels in the U.S. Population

CDC's Environmental Health Laboratory has been measuring lead levels in the population for many years. Lead poisoning can affect nearly every system in the body. It can cause learning disabilities, behavioral problems, and at very high levels, seizures, coma, and even death.

Our information about lead levels in the U.S. population resulted in the rapid removal of lead from gasoline in the United States and prompted research that showed similar relationships of gasoline lead to blood lead levels in other countries. This information also resulted in the removal of lead in gasoline in almost every industrialized nation.

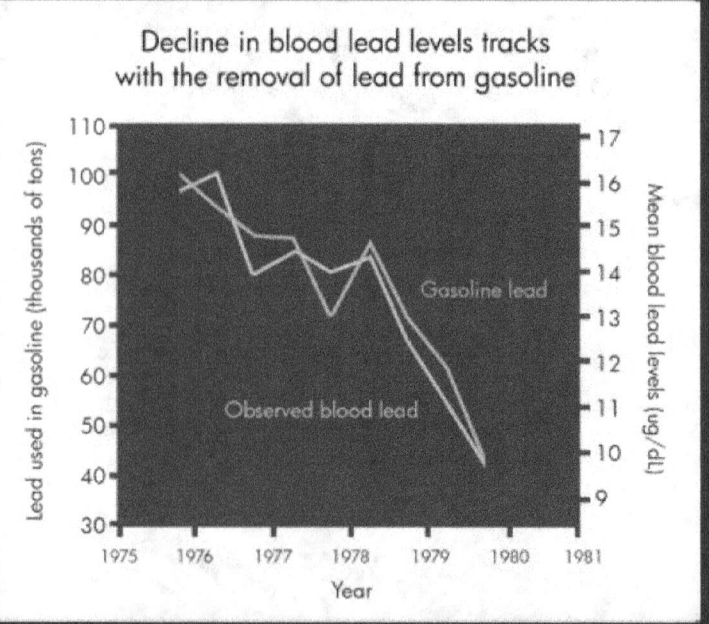

Decline in blood lead levels tracks with the removal of lead from gasoline

Investigating Exposure to Mercury at a Day Care Center

In 2006, CDC and the Agency for Toxic Substances and Disease Registry (ATSDR) were contacted about Kiddie Kollege Day Care Center in Franklin Township, New Jersey. The building that housed Kiddie Kollege was formerly a thermometer factory that used mercury and that subsequently was sold, renovated, and certified for use as a daycare center. Concerns surfaced about the past use of the building and potential impact of that use on the health of the children and staff. CDC's Environmental Health Laboratory analyzed urine samples from Kiddie Kollege staff and students for possible mercury exposure. Previous nationally representative data on human mercury urinary levels published in CDC's *Third National Report on Human Exposure to Environmental Chemicals* were used as reference levels to which levels in the children and staff were compared.

Understanding the employees' and parents' concern about the need for further testing, CDC and ATSDR provided additional assistance to the New Jersey Department of Health and Senior Services (NJDHSS) for several rounds of follow-up testing. As a further measure, the NJDHSS and ATSDR reviewed the medical records of any child or adult who attended or worked at the daycare center to assess whether an individual's past or present medical conditions were consistent with mercury poisoning. After each review, if further medical follow-up was indicated, the child's parents and physician were referred to the Pediatric Environmental Health Specialty Unit at the Mt. Sinai School of Medicine in New York or to the Robert Wood Johnson School of Medicine in New Jersey for additional testing or evaluation.

Most children had levels of mercury in their urine that were not unusual and that were consistent with national reference values. Some children had urine levels of mercury that were slightly higher than national reference values, and these children underwent several subsequent rounds of testing. The low urine levels of mercury in these children and the slight decline of the levels over time were consistent with background sources (i.e., diet or dental fillings) and a contribution from the daycare center exposure. No urine levels of mercury were in a range known to be toxic. Our laboratory tested a total of 189 urine samples during this investigation.

DID YOU KNOW?

In 2006, ESA Biosciences, a clinical diagnostics company, unveiled a portable blood lead testing instrument, LeadCare II, which was developed in consultation with IRAT. This instrument provides results rapidly—within 3 minutes—so that in one visit, a child with an elevated blood lead level can be identified and appropriate interventions initiated right away. In the past, at least two visits to a public health clinic or a doctor's office were required for obtaining results, and some families did not return for follow up.

Nutritional Biomarkers Branch

The Nutritional Biomarkers Branch (NBB) conducts research, develops methods, and analyzes essential nutrients (e.g., vitamins), nonessential nutrients (e.g., certain fatty acids), and bioactive dietary compounds (e.g., phytoestrogens and aflatoxin) that are responsible for changes in health status. For many of these essential nutrients and dietary compounds, the Branch produces population-based exposure levels segmented by age, sex, and race or ethnicity. This information is available in CDC's *National Report on Selected Biochemical Indicators of Diet and Nutrition in the U.S. Population,* the first national report to publish, in a single document, information about concentrations of 27 dietary and nutritional indicators found in the blood or urine of the U.S. population. This report establishes population reference ranges that can be used by physicians, clinicians, scientists, and health officials to determine whether a person or group of people has an unusually high or low level of a dietary or nutritional indicator. The report will also help determine whether the nutritional status of special population groups, such as minorities and potentially vulnerable groups (e.g., children, women of childbearing age, elderly), needs improvement. In addition, this nutrition report will assess the effectiveness of public health efforts to improve the diet and nutritional status of Americans.

The Branch also collaborates with academic, federal, and international partners on epidemiologic studies, intervention trials, and emergency response investigations. These efforts assess people's nutritional status or exposure to dietary compounds and their relation to health and disease.

Vitamin A Laboratory-External Quality Assurance

NBB operates the Vitamin A Laboratory-External Quality Assurance Program or VITAL-EQA. This program is designed to provide labs that measure micronutrients in serum with an independent assessment of their analytical performance. The program confirms the quality of the participating laboratories' analyses. Participation in VITAL-EQA is voluntary and free of charge. Results are not used for accreditation or certification but rather to help laboratories make sure they are performing their analyses correctly.

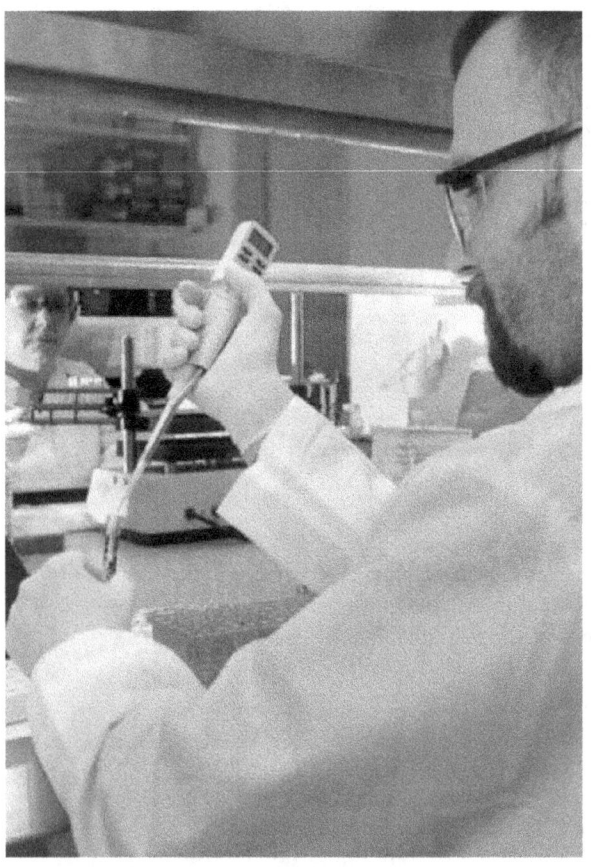

NBB is home to a total of 22 staff, including 1 physician, 9 people with Ph.D.s, 3 people with M.S. degrees, and 9 people with B.S. or other degrees.

DID YOU KNOW?

The Nutritional Biomarkers Branch reports nearly 100,000 test results every year for a wide range of nutritional biomarkers measured in NHANES and other studies. In collaboration with CDC colleagues, NBB also plays an active role in reducing the worldwide burden of micronutrient deficiencies, such as deficiencies in iodine, iron, or vitamin A. Billions of people around the world are affected by micronutrient deficiencies that may cause birth defects, learning disabilities, mental retardation, reduced immunity, blindness, poor work capacity, or premature death. The laboratory provides training and technical assistance on questions about laboratory analysis and field logistics in preparation for national nutrition surveys in countries such as Argentina, the Dominican Republic, Iraq, Kenya, Laos, Malawi, Nicaragua, Oman, Papua New Guinea, South Africa, Ukraine, Uzbekistan, Yemen, and Zambia.

Measuring Folate Levels in People

NBB has provided long-time scientific leadership in the area of folate nutrition by 1) developing state-of-the-art methods for measuring levels of folate in people's blood and serum, 2) helping with efforts to standardize clinical methods for measuring folate in people, and 3) measuring folate levels in the U.S. population over many years. Knowing folate levels is important because women of childbearing age with low folate levels are at risk of giving birth to a baby with birth defects of the brain or spine. In 1996, the U.S. Food and Drug Administration mandated that all enriched cereal-grain products be fortified with folic acid by January 1998. Food fortification was determined to be the best strategy for increasing blood folate levels since the critical period for adequate folic acid intake is in the first weeks of pregnancy, before most women know they are pregnant and begin taking prenatal vitamins.

NBB measured levels of folate in people who participated in NHANES, which collects data on the health of people living in the United States through interviews, direct physical examinations, and laboratory tests. The laboratory found that serum folate levels have nearly tripled in the U.S. population since 1998.

Investigating a Public Health Emergency in Kenya

In May 2004, health officials reported an outbreak of jaundice in two districts in Kenya with a high case-fatality rate. This outbreak was caused by widespread contamination with aflatoxin of locally grown maize. Aflatoxin is the name given to any group of toxic compounds produced by certain molds. In this case, fungus had grown on the grain and produced a toxin. NBB played a major role in the success of this investigation through its use of sophisticated laboratory techniques to provide the best possible evidence that people in the two affected areas had been exposed to aflatoxin. These techniques, which had never before been used in an outbreak, directly measured toxins in blood samples from people who had eaten the maize. As a result, officials had a much more accurate assessment of exposure than would have been possible by estimating exposure on the basis of measuring toxins in food samples. These efforts not only helped stem the outbreak in Kenya but will also be key to developing strategies for preventing future tragedies caused by people's exposure to aflatoxin.

Newborn Screening and Molecular Biology Branch

Newborn Screening

The Newborn Screening and Molecular Biology Branch (NSMBB) has the only laboratory in the world devoted to ensuring the accuracy of newborn screening tests in every state and more than 57 countries. Newborn screening is a vital public health program that tests babies for congenital disorders that are not apparent at birth. The Newborn Screening Quality Assurance Program (NSQAP) develops analytical methods to measure substances in dried blood spots (DBSs) and produces certified DBS quality-control and reference materials for newborn screening tests. Because of NSQAP, parents and doctors in the United States and worldwide can trust the results of newborn screening tests. In 2005, NSQAP received the Charles C. Shepard Science Award for Outstanding Scientific Contribution to Public Health. The award was given in recognition of the program's mission, which helps ensure that screening done by all state public health laboratories detects metabolic or genetic disorders and does not miss such cases.

In 2005, the Branch launched the Newborn Screening Translation Research Initiative (NSTRI) as an ongoing collaboration with the CDC Foundation. Working with corporate, academic, and foundation partners, NSTRI assures the quality of research methods during pilot studies and into routine screening. Since its inception, NSTRI has developed laboratory projects focusing on a variety of disorders, including lysosomal storage disorders and severe combined immune deficiency.

Genetic Studies

NSMBB also conducts research, develops methods, and performs analyses by using complex, state-of-the-art molecular techniques for identifying genetic risk factors of public health importance. A variety of diseases have been or are currently the topic of study, including type 1 diabetes and kidney disease, asthma, type 2 diabetes, ischemic stroke, an iron-overload disease known as hemochromatosis, birth defects, and acute lymphoblastic leukemia. The Branch is also evaluating variations in genes that may influence how people smoke and metabolize the cancer-causing agents in cigarettes.

The NSMBB staff are establishing DNA repositories and performing quality control for CDC's NHANES, the Genetics of Kidneys in Diabetes Study, the National Birth Defects Prevention Study, and the Stroke Prevention in Young Women Study. Quality control of the DNA repositories is essential because these collections are used in studies by CDC and researchers worldwide to better understand how genetics relate to disease and health. NSMBB is also collaborating with the CDC Foundation and with CDC partners to launch the Beyond Gene Discovery Initiative, which will use the NHANES DNA bank to assess population genetic variation in the United States in relation to health and disease.

Standardizing Diabetes Measurements

The Branch also develops materials and methods to improve measurements of autoantibodies that are predictive of type 1 diabetes. These are the most sensitive and meaningful measures for predicting this disease. Historically, autoantibody measures have been variable among laboratories; therefore, the Branch established the Diabetes Autoantibody Standardization Program (DASP) in collaboration with the Immunology of Diabetes Society. The goals of DASP are to improve laboratory methods, evaluate laboratory performance, support the development of sensitive and specific measurement technologies, and develop reference methods. Currently, 48 key laboratories from 19 countries participate in DASP.

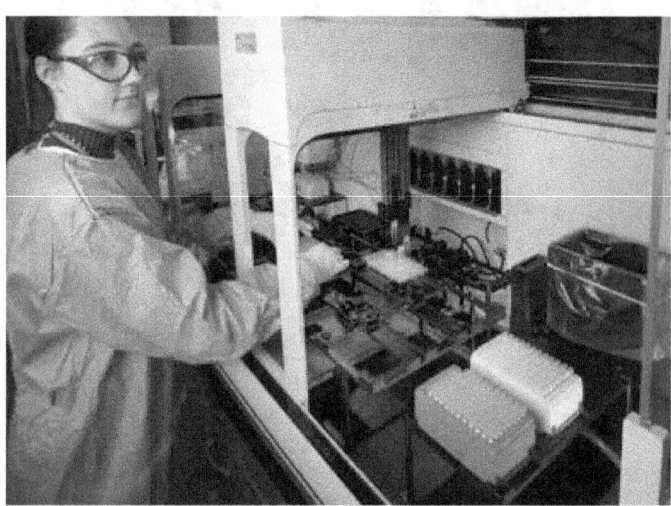

NSMBB is home to a total of 41 staff, including 14 people with Ph.D.s, 5 people with M.S. degrees, and 23 people with B.S. or other degrees.

Helping States Develop Newborn Screening Tests

Severe combined immune deficiency (SCID) is caused by genetic mutations that result in a failure of the immune system to develop normally. Sometimes referred to as "Bubble Boy Disease," SCID is characterized by an inability to resist infections. Without early diagnosis and treatment, babies with SCID usually die within a year of birth. Newborns with SCID are protected by maternal antibodies for the first few weeks following birth, but after that they develop severe recurrent infections. Babies who are identified with SCID before they become infected can be treated by bone marrow transplantation. Early treatment is essential to reduce costly hospitalizations and produce the best chance for a successful outcome. Since the development of a test for SCID that can be used on dried blood spots, CDC's Newborn Screening Translation Research Initiative has worked with partners from public health, foundations, and universities to pursue population-based pilot studies so that newborn screening for SCID can be evaluated by state public health newborn screening programs. With Congressional funding, CDC is providing cooperative agreements that will support states in these efforts.

Investigating a Cluster of Cases of Leukemia among Children in Nevada

CDC worked with the Nevada State Health Division (NSHD) to conduct a study to identify environmental exposures in the Churchill County, Nevada community, where 15 cases of childhood leukemia had been diagnosed from 1997–2002. CDC found that people living in Churchill County had higher levels of some metals in their urine than the U.S. population. Two of these metals were tungsten and arsenic. CDC also found that the levels of these metals in families in which a child had leukemia (case families) were not different from the levels found in families in which no children had leukemia (comparison families).

In collaboration with St. Jude Children's Research Hospital, CDC tried to find out if there were genetic similarities or differences between case children and comparison children so that these factors could be examined. CDC found a genetic difference, or variant, in a gene that contains instructions for making special proteins called sulfite oxidase enzymes. Without this and other enzymes, the body may not be able to detoxify chemicals (i.e., to change a harmful substance into a safer form).

It is not clear how inactivation of this enzyme might relate to childhood leukemia. CDC found no evidence that a genetic factor caused the cases of leukemia in Churchill County. It is also not possible to say that the presence of this genetic variant can predict who will get leukemia. That's because all of the case children and 40 percent of comparison children had this variant, and many of these children were also exposed to tungsten. More research may be useful to find out if having the gene variant increases children's susceptibility to leukemia and what other factors, if any, must exist to cause leukemia. This study is an important first step in answering these and other difficult questions about childhood leukemia in Churchill County.

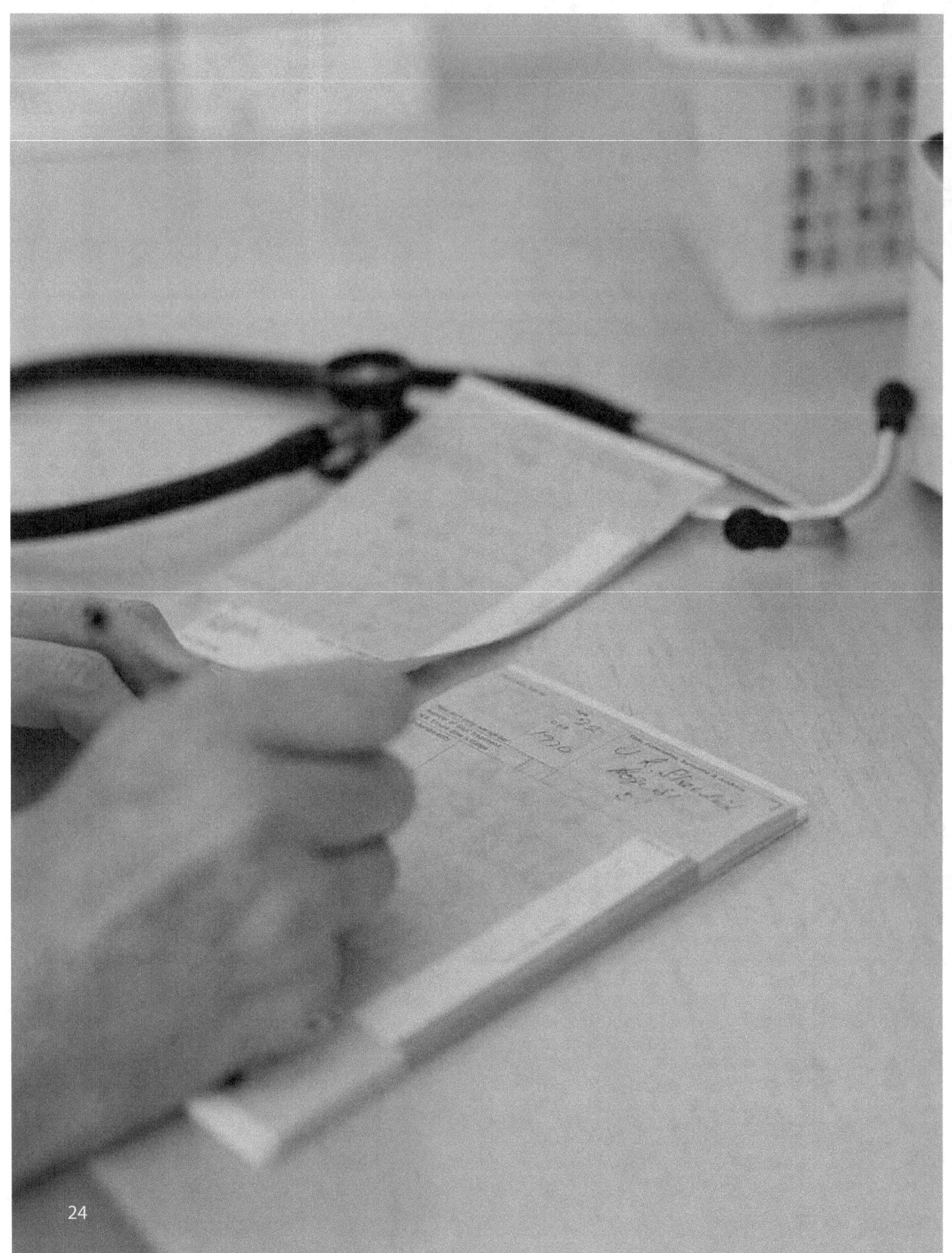

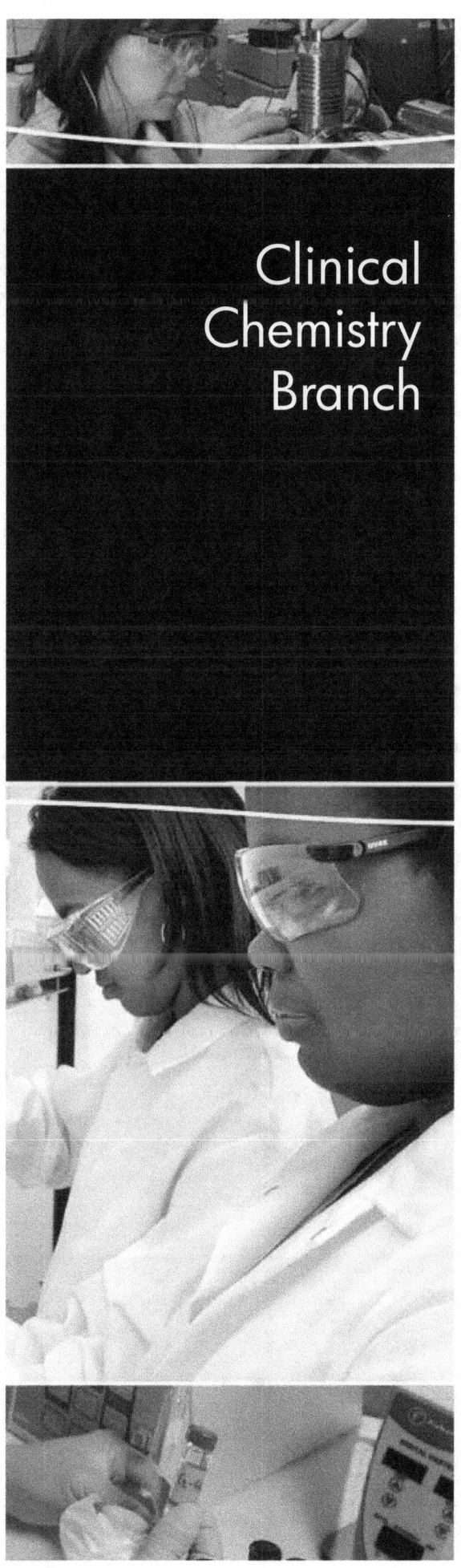

Clinical Chemistry Branch

The Clinical Chemistry Branch (CCB) develops and improves methods for assessing disease status associated with and the risk for selected chronic diseases, including cardiovascular disease, diabetes, and hormonal disorders. The Branch serves as a reference laboratory for these measurements by providing standards that are used as accuracy points by other laboratories around the world. By operating as the world reference laboratory for certain chronic disease measurements, the Branch ensures that measurements from different research studies can be compared. Ultimately, the findings from these research studies are translated into clinical applications, thus enabling doctors to tailor treatment for their patients.

Cardiovascular Disease

CCB serves as the world reference laboratory for measuring cholesterol, triglycerides, high-density lipoproteins, and low-density lipoprotein cholesterol. In collaboration with the National Heart, Lung, and Blood Institute (NHLBI), the Branch operates the Lipid Standardization Program to ensure the quality of about 35 million cholesterol measurements done annually in the United States alone. With accurate measurements, doctors can diagnose and properly treat people with high cholesterol levels, thus reducing illness and death associated with cardiovascular disease. LSP also standardizes lipid measurements for 100 laboratories worldwide that are involved in ongoing epidemiologic studies or clinical trials investigating risk factors for cardiovascular disease. By serving as the accuracy base for lipids and lipoproteins, LSP ensures the comparability of results obtained from these investigations.

Diabetes

CCB serves as a reference laboratory for measuring hemoglobin A1c. The A1c test measures the amount of glucose in a person's blood over a 2- to 3-month period and is a tool to monitor diabetes. The Branch supports the efforts of the International Federation of Clinical Chemistry and Laboratory Medicine to standardize glycated hemoglobin A1c, and it provides support to the National Glycohemoglobin Standardization Program. These activities improve the accuracy of A1c measurements nationally and internationally. CCB also collaborates with the National Kidney Disease Education Program to standardize and improve measurements used to assess chronic kidney disease.

Measuring Environmental Chemicals in People

CCB researchers also develop and use analytical methods to assess people's exposure to such chemicals as acrylamide, *trans*-fatty acids, and ethylene oxide. These methods will allow CDC to obtain initial data about people's use of and potential exposure to these chemicals. These data will be published in CDC reports about the U.S. population's exposure to chemicals in the environment as well as in reports of the levels of biochemical indicators of diet and nutrition in the population.

Branch scientists also regularly provide statistical consultation services in research, study design, data analysis, reporting, and quality-control development for laboratory investigations and environmental health studies to numerous international, federal, state, and local agencies and organizations.

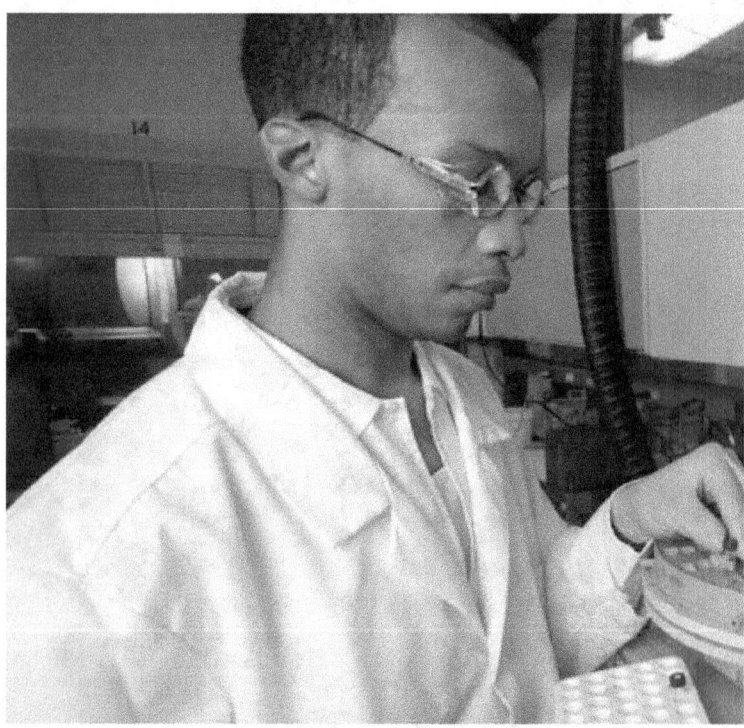

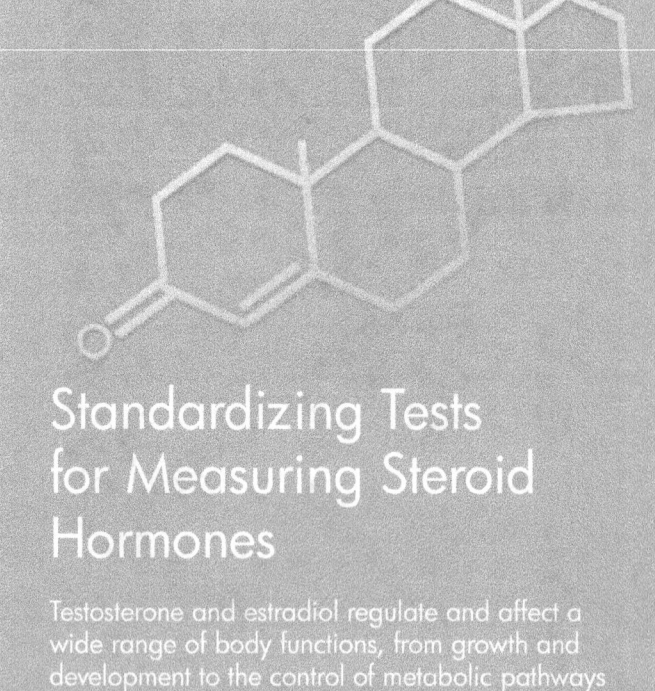

Standardizing Tests for Measuring Steroid Hormones

Testosterone and estradiol regulate and affect a wide range of body functions, from growth and development to the control of metabolic pathways and the use and storage of energy. Testing for these compounds has become an important tool in the research and management of diseases and disorders such as cancer or polycystic ovary syndrome. The need for standardization of these tests has been stated by the clinical and research community. CCB responded to this need by starting a project to standardize these tests, thus improving prevention and control of steroid hormone-related diseases and disorders.

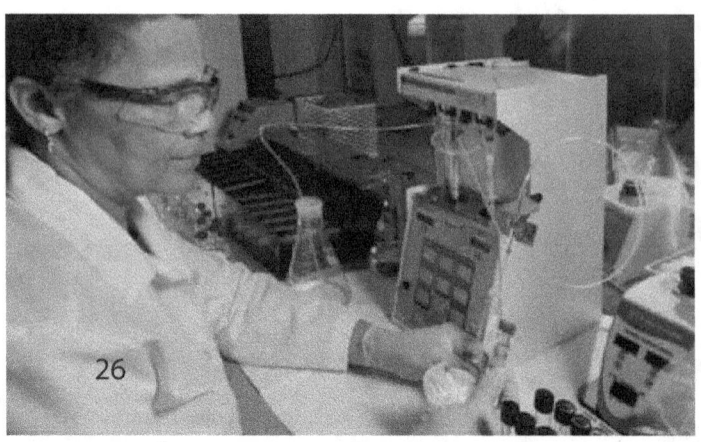

CCB has 28 staff members, including 1 physician, 9 people with Ph.D.s, 3 people with M.S. degrees, and 15 people with B.S. or other degrees.

A Method for Determining People's Exposure to Acrylamide

CCB has developed a method for measuring the chemical acrylamide in blood. Acrylamide forms in certain foods when they are cooked at high temperatures. Studies have shown it causes cancer in animals and is suspected to cause cancer in people. The levels of acrylamide in food are much lower than the amount known to cause disease; however, little is known about how acrylamide in foods affects the levels of acrylamide measured in blood. CCB is conducting a study that will assess the impact of acrylamide exposure from food. This study will aid the Food and Drug Administration in assessing the risks associated with exposure to acrylamide in food.

DID YOU KNOW?

CCB operates the Cholesterol Reference Method Laboratory Network (CRMLN), which certifies manufacturers of clinical diagnostic products for measuring cholesterol. CRMLN conducts tests that compare the results of the manufacturers' diagnostic assays with CDC's reference methods for lipids and lipoproteins. Through this process, manufacturers can adjust their analytical systems to assure physicians that the test results are accurate.

Healthy People in Every Stage of Life

The Division of Laboratory Sciences of CDC's National Center for Environmental Health protects people in every stage of life—from before birth, during the newborn months, through childhood and adolescence, and into adulthood.

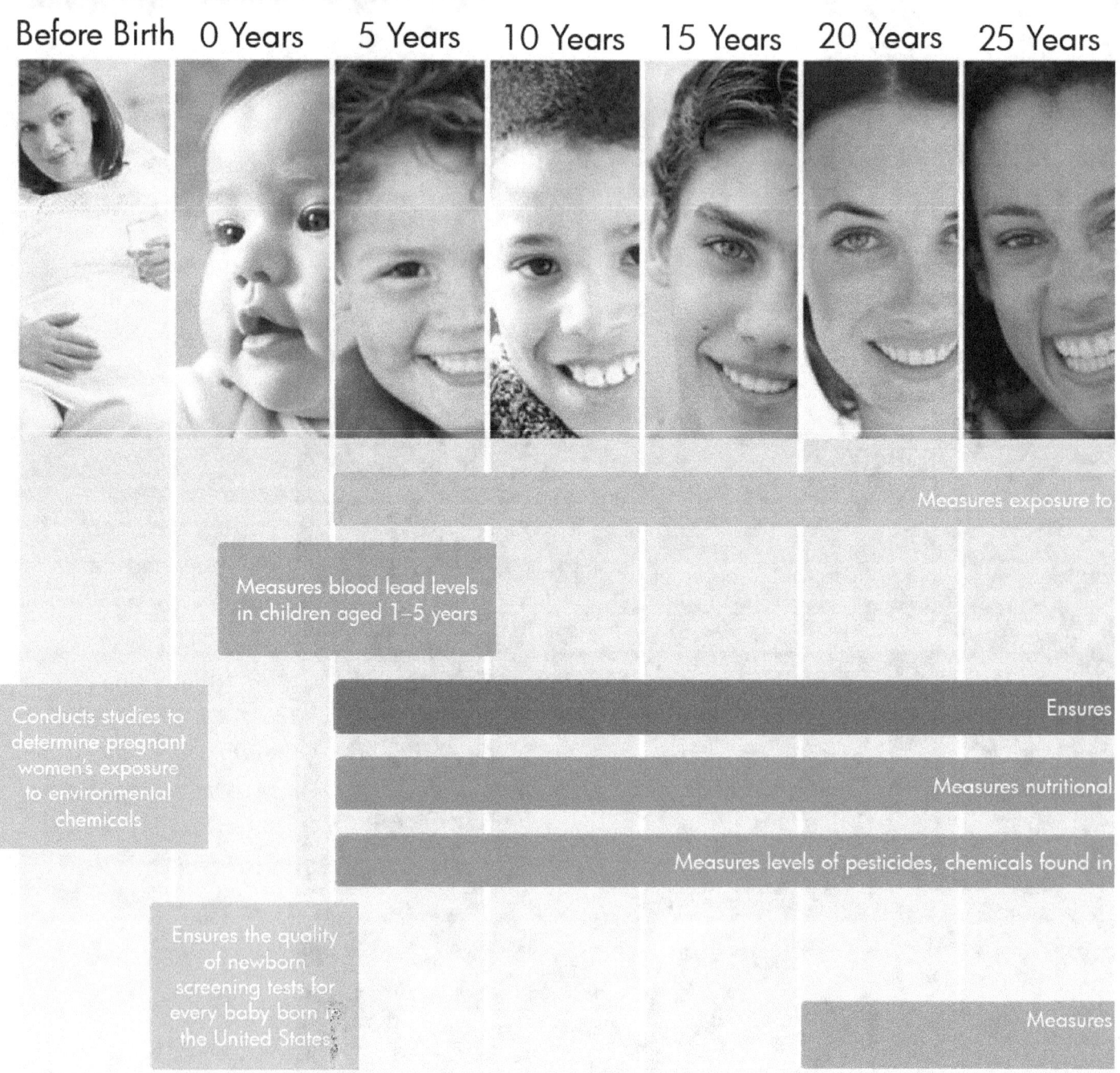

Before Birth	0 Years	5 Years	10 Years	15 Years	20 Years	25 Years

Measures exposure to

Measures blood lead levels in children aged 1–5 years

Conducts studies to determine pregnant women's exposure to environmental chemicals

Ensures

Measures nutritional

Measures levels of pesticides, chemicals found in

Ensures the quality of newborn screening tests for every baby born in the United States

Measures

30 Years   35 Years   40 Years   45 Years   50 Years   55 Years   60+ Years

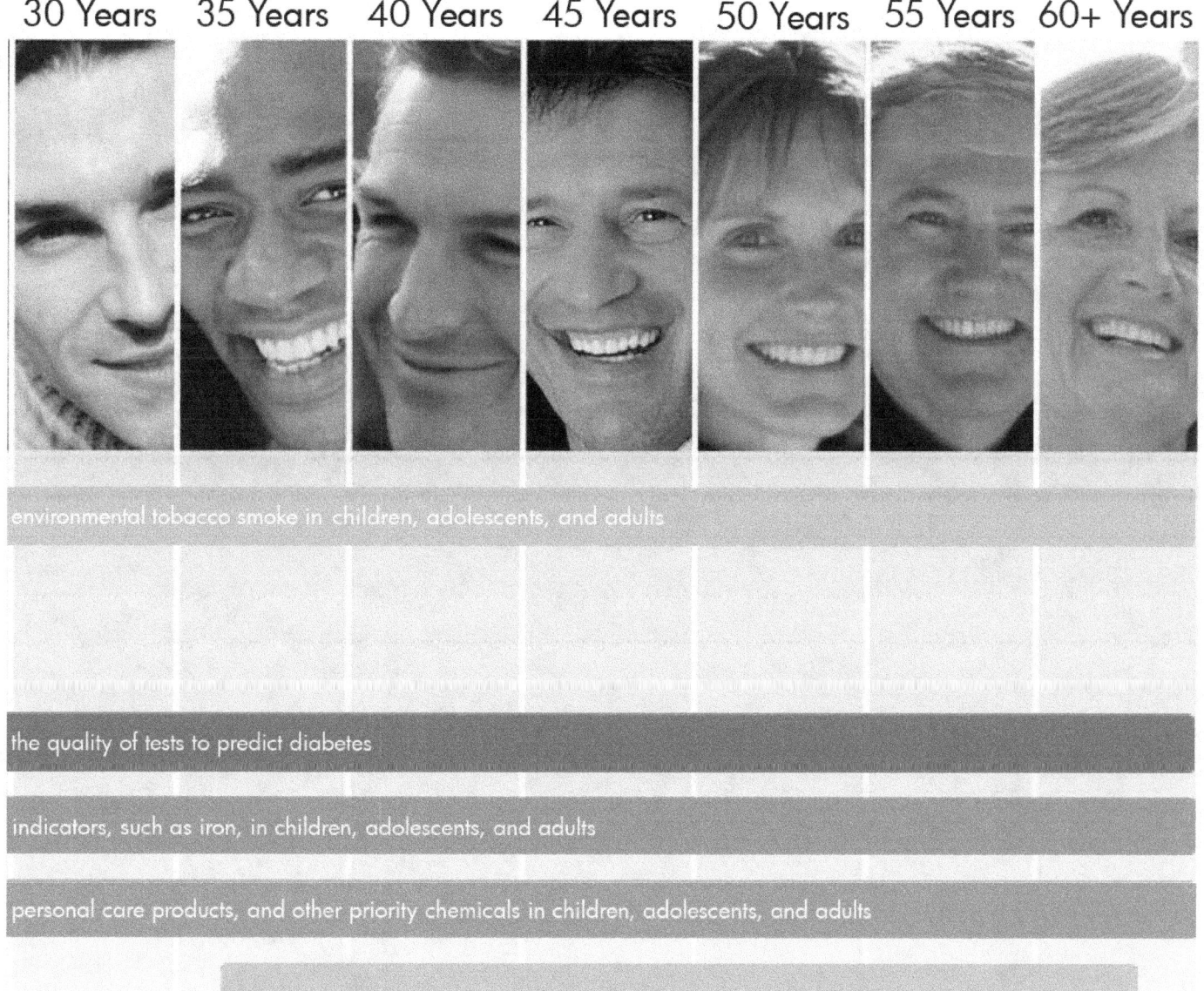

environmental tobacco smoke in children, adolescents, and adults

the quality of tests to predict diabetes

indicators, such as iron, in children, adolescents, and adults

personal care products, and other priority chemicals in children, adolescents, and adults

Provides the world reference for cholesterol measurements

nutritional indicators, such as folate levels,
in women of child-bearing age

29

Centers for Disease Control and Prevention
National Center for Environmental Health
Division of Laboratory Sciences
Mail Stop F-20
4770 Buford Highway, NE
Atlanta, Georgia 30341-3724
Telephone (toll-free): 1-800-CDC-Info (1-800-232-4636)
Email: CDCINFO@cdc.gov
Web site: www.cdc.gov/nceh/dls

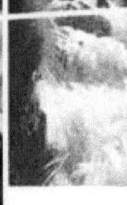
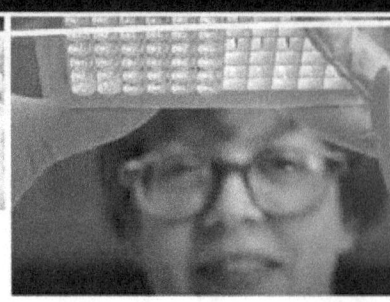